THE CHANGING EARTH

DOUGAL DIXON

Wayland

Young Geographer

The Changing Earth
Food and Farming
Natural Resources
Protecting the Planet
Settlements
The World's Weather

DDS-93-325

Editor: Sarah Doughty
Designer: Mark Whitchurch
Consultant: Dr Tony Binns, geography lecturer at Sussex
University

Front cover picture: Rock formations in Arches National Park,
Utah, USA.
Back cover picture: The Baltoro glacier in Pakistan.
Frontispiece: A view of Mount St. Helens, USA, after the
volcano exploded in 1980.

First published in 1992 by
Wayland (Publishers) Ltd
61 Western Road, Hove
East Sussex BN3 1JD, England

© Copyright 1992 Wayland (Publishers) Ltd

British Library Cataloguing in Publication Data
Dixon, Dougal
The Changing Earth. – (Young Geographer)
I. Title. II. Series
551.4

ISBN 0 7502 0385 4

Typeset by Type Study, Scarborough, England
Printed in Italy by Rotolito Lombarda S.p.A
Bound in France by A.G.M.

National Curriculum Attainment Targets

This book is most directly relevant to the following
Attainment Targets in the Geography National Curriculum
at Key Stage 2. The information can help in the following
ways:

Attainment Target 1 (Geographical skills) Developing a
geographical vocabulary to talk about landscapes;
identifying familiar features on photographs and pictures.

Attainment Target 3 (Physical geography) Identifying and
describing landscape features (such as valleys, beaches and
rivers); earth movements and the nature and effects of
volcanic eruptions; water in its different forms; what
happens to rainwater when it reaches the ground; the parts
of a river system; erosion, transportation and deposition of
materials by rivers, waves, winds and glaciers; physical and
chemical weathering and erosion.

Attainment Target 5 (Environmental geography) Showing
the ways people can damage the environment through
activities like farming.

Contents

All the words that are in **bold** appear in the glossary on page 30.

Introduction

Can you remember when your neighbourhood looked different from how it looks now? You can probably recall a new house being built, or a new fence erected. You may even remember when a new road was laid down through the countryside or the town.

Human beings have always been changing the scenery like this. We can do it on a small scale, or on a very large scale. We can change the landscape by damming rivers and forming lakes, by clearing forests to make farmland, or by building sea walls to make dry land where there was once sea.

These changes take place in a few years or over a few decades. You can often see the difference when you go back to a place after being away for a time.

We can change our landscape dramatically when we build houses and roads. Yet these changes are quite small compared with the changes brought about by nature.

The power of nature, as in the crashing of waves against a shoreline, can wear away the rocks and change the landscape completely over the years.

Nature makes far greater changes than this. The sea washes cliffs away. New sandbanks become islands and are then joined to the mainland. Mountains are thrown up, and then worn away to plains. Rivers wash the soil away to the sea.

The great changes brought about by nature usually take much longer than those brought about by humans, who have only changed the landscape over the last few thousand years. Nature's changes take place so slowly that you cannot see them happening. But it has been working and reworking the landscape ever since the world formed, over four and a half thousand million years ago.

Mountains

Some of our mountains are so rugged that they look as if they might have been there for all time. But in fact, they may be quite new. Many mountain ranges have formed since the days of the dinosaurs. The highest and most rugged mountains of the world such as the Himalayas and the Alps, are less than fifty million years old and are actually the youngest.

Older mountains have stood for long enough to have been worn away by time and rotted by the weather until they are gentle and rounded, like the highlands of Scotland or the Appalachians of eastern North America.

The Himalayas – the greatest mountain range on Earth – are geologically quite recent.

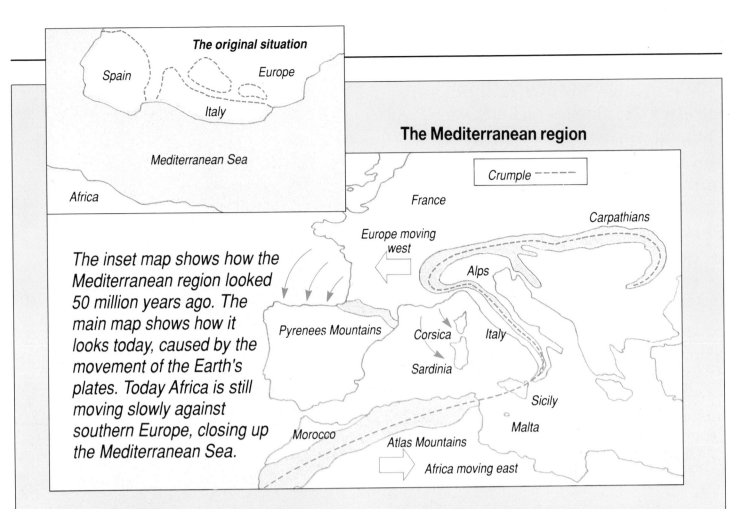

The original situation

Spain

Europe

Italy

Mediterranean Sea

Africa

The Mediterranean region

Crumple ------

France

Carpathians

Europe moving west

Alps

The inset map shows how the Mediterranean region looked 50 million years ago. The main map shows how it looks today, caused by the movement of the Earth's plates. Today Africa is still moving slowly against southern Europe, closing up the Mediterranean Sea.

Pyrenees Mountains

Corsica

Italy

Sardinia

Sicily

Morocco

Malta

Atlas Mountains

Africa moving east

The mountains along the Mediterranean Sea are all twisted. This is because Africa has been moving eastwards past Europe.

As these two continents slid past one another, the mountains on the coast twisted into a great S-shape – from the Atlas Mountains of Morocco and Algeria, across the islands of Malta and Sicily, and the 'boot' shape of Italy, round the vast sweep of the Alps and then finally curving around into eastern Europe with the Carpathians.

At the same time Spain and Portugal were pulled from their position in what is now the Bay of Biscay and swung into the place they are now. The islands of Corsica and Sardinia were plucked from the bays of southern France.

It is the movement of the **tectonic plates** which make up part of the Earth's **crust** that push up the mountains. The plates move about at the surface of the Earth, usually at a rate of a few centimetres per year. Sometimes the plates collide, or move apart forming mountain ranges. There are several ways this can happen.

When one **continent** collides with another the coastal mountains of each continent will form into a single vast mountain range. Fifty million years ago India collided with the mainland of Asia and the Himalayas were thrown up in between.

These great mountain chains of the world are called **fold mountains**, because the rocks have been folded up by Earth movements. We can actually see that this has happened by looking at some of the bent and twisted rocks on mountainsides.

There are other types of mountains too – called **block mountains**. These form when continents are being pulled apart rather than squeezed together.

The effects of the forces that produce mountains can be seen where beds of rock have been twisted and folded.

Fold Mountains ▶

The greatest mountain ranges are formed where two plates of the Earth's surface have been pushed together. The rocks in between are crumpled up along the join between the two. This action produces the fold mountains.

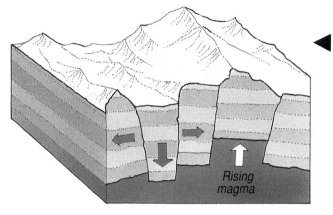

◀ Block Mountains

Where the Earth's crust is pulled apart it cracks and splits into separate blocks. Some of these blocks drop down leaving others upstanding. These upstanding blocks form the block mountains.

Dome Mountains ▶

Molten material in the Earth's crust may surge upwards, forcing up the solid rocks above. On the surface the bulge produced gives rise to mountains called dome mountains.

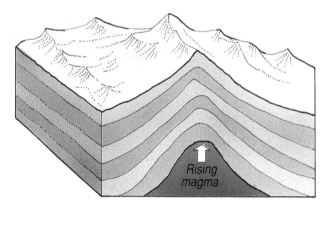

The plates below the Earth stretch and pull at the rocks so they crack into blocks. These blocks move up and down in relation to each other along cracks called **faults**.

Other mountains are called **dome mountains**. They are formed by molten material welling up from below the Earth's surface. **Magma** seeps into existing layers of rock, arching up underneath the overlying rock layers.

After this activity has taken place, the movement usually stops and the rock layers begin to wear away revealing the magma dome.

Volcanoes

The heat beneath the crust of the Earth can melt the deep rocks and blast them to the surface. When this happens, rocks and **lava** spill in all directions as a **volcanic** eruption. As volcanoes erupt, they build up into a hill or mountain. Two main types of volcano can be formed.

One type of volcano forms as molten material from deep below the Earth's crust is forced upwards.

This produces hot fountains of glowing lava which then flows over the ground. Lava forms rivers of fire that can flow for long distances before cooling and forming solid rock. The lava hardens and builds up to form a volcanic mountain.

Because this kind of lava flows for a long way, the volcanic mountain that forms tends to be very broad and flat.

Molten rock spilling out at the surface of the Earth can form red-hot rivers of lava. They can flow a long way before cooling and solidifying.

Hawaii

Far out in the Pacific Ocean, the islands of Hawaii lie over an area called a **hot spot**. Molten material from inside the Earth makes its way through the crust above and forms a volcanic island. But as the Earth's crust is moving all the time, the volcano that has built up is carried away from the hot spot. The old volcano then begins to erode away, while a fresh volcano erupts beside it at the place that now lies over the hot spot.

The Hawaiian island chain is made up of one big island with active volcanoes, and a string of smaller islands – each one older than the next, the further away it is from the main island. What is now left of these ancient underwater volcanoes stretches half-way across the Pacific Ocean.

Above *The very fluid lava of the Hawaiian volcanoes spreads out forming low flat volcanic mountains.*

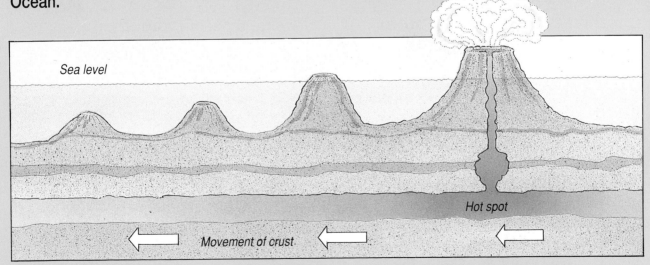

Sea level

Hot spot

Movement of crust

As the Earth's surface moves over a hot spot, a chain of volcanic islands forms.

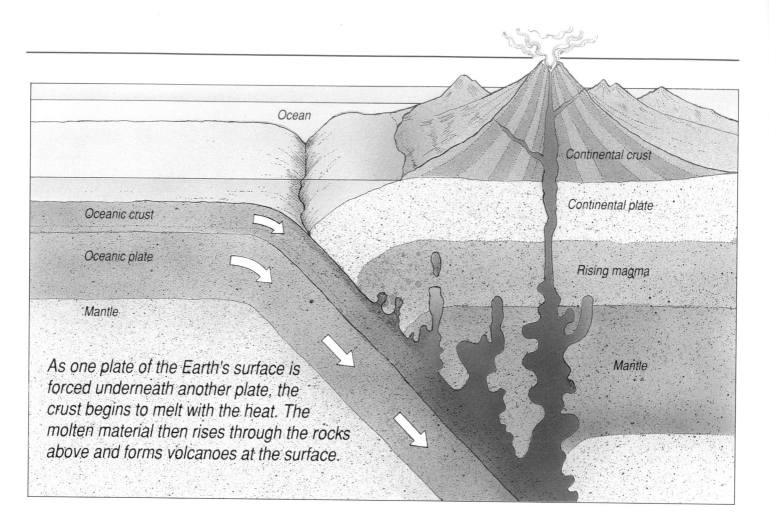

As one plate of the Earth's surface is forced underneath another plate, the crust begins to melt with the heat. The molten material then rises through the rocks above and forms volcanoes at the surface.

Labels in diagram: Ocean, Oceanic crust, Oceanic plate, Mantle, Continental crust, Continental plate, Rising magma, Mantle

The second type of volcano is produced when the material of the Earth's crust melts. This happens when there are movements of the Earth's surface plates – usually one plate collides with another, forcing one plate to dive underneath the other. This can make part of the Earth's crust sink.

The rocks of the crust melt and are forced upwards to the surface to form volcanoes. The lava from this type of volcano is very sticky and does not flow far. Because of this, the volcanoes are tall and conical.

Often the magma becomes solid before it reaches the surface, and blocks up the tube through which it flows. The pressure building up below can then cause a disastrous explosion.

Magma rising from below the ground has formed volcanoes in mountain ranges such as the Andes. Over millions of years, partly melted magma has been slowly forced through the rocks and hardened. Today these rocks have solidified and have been eroded into the peaks of the Andes mountains.

Mount Pinatubo, Philippines

The Philippines are made up of a series of volcanic islands which have been formed by the melting of the Earth's crust. This happens because Asia is pushing its way out into the Pacific Ocean.

Now and again these volcanoes erupt suddenly. After 600 years asleep, Mount Pinatubo exploded in April 1991, and then continued to erupt for months after that.

The explosions were so great that volcanic ash rained down on nearby villages. Within three weeks of the biggest explosion, the gas and dust had circled the world.

Heavy rains then softened the loose ash on the steep volcanic slopes, and mudslides swept into the local villages. Volcanoes like this are very dangerous.

A flow of mud and ash destroys villages near Mount Pinatubo.

Coasts

Visit the seaside on a stormy day. You will be impressed by the power of the thundering waves that crash against the sea walls. So it is hardly surprising that the ocean is continually chewing away at the edge of the land, smashing the cliffs to rubble and grinding the rocks to sand. Where the land juts out into the sea, the waves sweep round and attack the **headland** from each side.

As the waves curl over and smash into cliffs, the cracks and hollows in the rock are forced open. Pieces are plucked out. The holes become bigger and sea caves form. As the waves surge in and out of a cave, air pressure can build up and burst a hole in the roof. The result is a **blowhole**, where spray is blasted upwards every time a wave comes in.

Cliffs, such as these along the north-west coast of Australia, are eroded by the waves, leaving seastacks and pinnacles that eventually crumble into the sea.

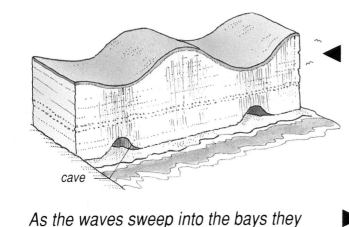

cave

As the waves sweep into the bays they curve round and crash into the sides of the headlands. Any cracks or hollows in the headlands are opened up by the force of the battering. They widen into sea caves.

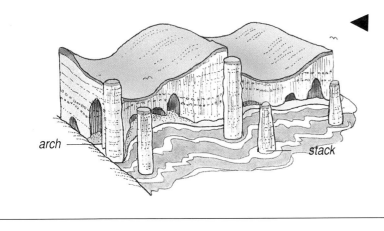

arch

stack

Cliff Erosion

When a cliff is eroded by the waves the softest rocks are worn away first. Bays are cut back into the soft rock, leaving headlands of the harder rock at each side.

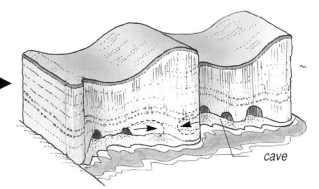

cave

With the continuing pounding of the waves the sea caves become larger and erode the headland into a series of natural arches. These eventually collapse to form seastacks. In this way the cliff is gradually worn back.

When caves at each side of a headland meet in the middle they form a natural arch. Eventually the arch will become so wide that the roof falls in. This leaves the seaward part as a single **seastack**, which over time is eroded away.

At a coastal cliff face, the waves attack at the base. This undercuts the cliff, and part of it collapses. This happens again and again so the cliff slowly goes back. It leaves a **wave-cut platform** – a shelf of rock at sea level. The broken rocks from the cliff are ground down by being rolled back and forth over the sea-bed and become rounded pebbles or shingle.

Eastern England

On the coast of Yorkshire there are cliffs of hard limestone and chalk. These jut out into the North Sea and are attacked by fierce waves. The headlands here are cut back as a series of caves, blowholes, arches and stacks.

The sea currents along the east coast of England move mostly southwards. The material from the eroding cliffs is carried in this direction. Many of the rivers that reach the sea along the east coast have long sand spits across their mouths. One of these lies at the mouth of the Humber, where a long curved hook of sand called Spurn Head reaches across the **estuary**.

Right Coastal erosion and deposition on Britain's eastern coast.

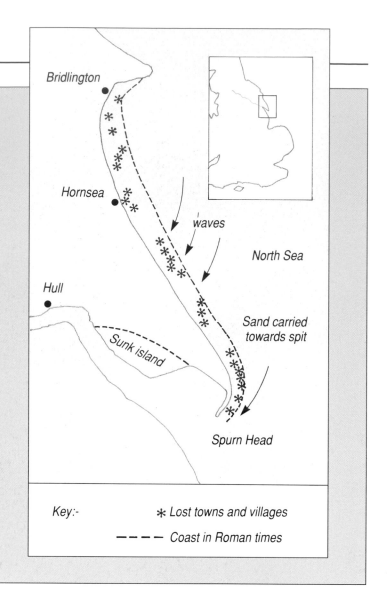

Key:-

* Lost towns and villages

\- - - - Coast in Roman times

The broken rocky material is swept along by the sea currents, often a great distance from the cliffs. When the waves and the currents are not so strong, the material settles on the sea-bed or along the shore. This produces a **beach**. In stormy weather a beach made of shingle will throw up a high bank called a **berm**. At quieter times of the year the beach may become more sandy.

As the sand is swept along the coast by the currents, it may reach a bay. The sand may then build out across the mouth of the bay making a sandbar. The water of the bay will then be cut off from the sea as a shallow **lagoon**.

When this happens at a river mouth, the river current will stop the bar from going right across the river mouth. The result is a curved ridge of sand called a sand **spit**.

Wave action produces a wave-cut platform at sea level.

Rivers

When rain falls on the hills the water seeps down through the soil and rocks. Eventually the soil and rocks are completely filled with water and will not hold any more. At certain places, such as on hill slopes, the water reaches the surface and seeps out as a spring.

This water may form a stream. The stream will merge with other streams, or **tributaries**, and grow into a river that runs splashing downwards in waterfalls and rapids. It cuts a deep V-shaped gorge as it goes and carries along rocks and gravel plucked from its bed.

The force of water tumbling down a mountain waterfall shows the erosive power of the upper section of a river.

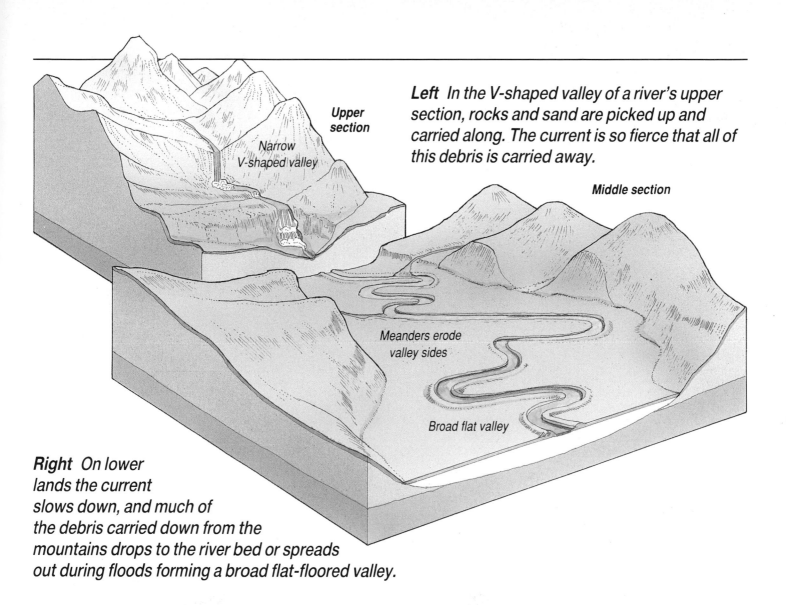

Upper section

Narrow V-shaped valley

Left In the V-shaped valley of a river's upper section, rocks and sand are picked up and carried along. The current is so fierce that all of this debris is carried away.

Middle section

Meanders erode valley sides

Broad flat valley

Right On lower lands the current slows down, and much of the debris carried down from the mountains drops to the river bed or spreads out during floods forming a broad flat-floored valley.

This is the upper section of a river. As the river flows out of the hilly areas on to the plains it enters its middle section. Here it runs more slowly, winding back and forth across a broad flat-bottomed valley. Much of the material brought down from the hills is dropped here, forming the flat valley floor.

The river follows a winding course across the valley making wide curves. The water flows more quickly round the outsides of the curves and cuts back the river bank. The slow waters on the insides of the curves means that sand and gravel are dropped here as beaches.

In this way each curve, called a **meander**, moves across the valley floor. When the river flows close to the valley wall it erodes the sides of the valley back into steep banks. In this way the valley becomes steadily broader.

In a river's lower section the water has no power left. All the rest of the debris carried down from the mountains is dropped and forms a floodplain.

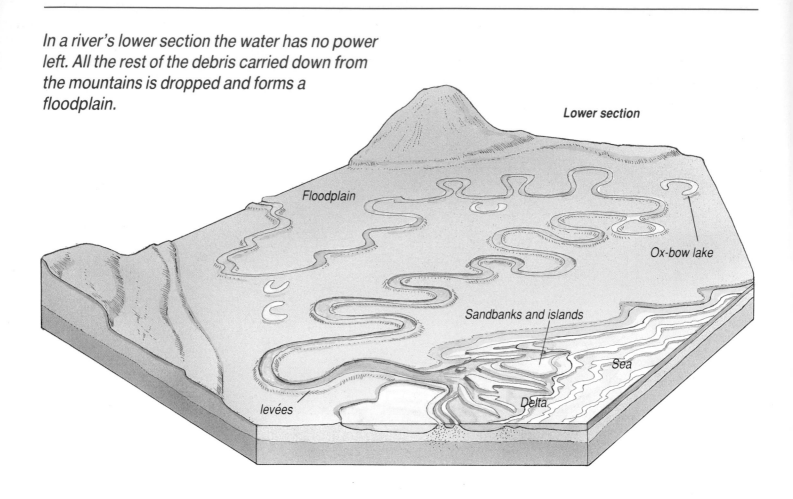

Lower section

Floodplain

Ox-bow lake

Sandbanks and islands

Sea

Delta

levées

By the time the river has reached its lower section it has left the mountains and hills far behind. It flows sluggishly over a flat plain called a floodplain. Here it flows so slowly that the river cannot erode any more. The rest of the material is dropped here, usually along the banks at times of flood.

As a result the banks build up, forming **levées**. Sometimes the river between the levées is at a higher level than the plain around them.

Whenever the river breaks through, it deposits material over a wide area. There are many meanders at this stage, and sometimes the curves are abandoned and left as lakes, called ox-bow lakes.

The river finally reaches the sea, and if the sea is a quiet bay the rest of the material is dropped here. Sandbanks and islands build up, and the river has to cut channels through them. This whole area is called a **delta**.

The Mississippi

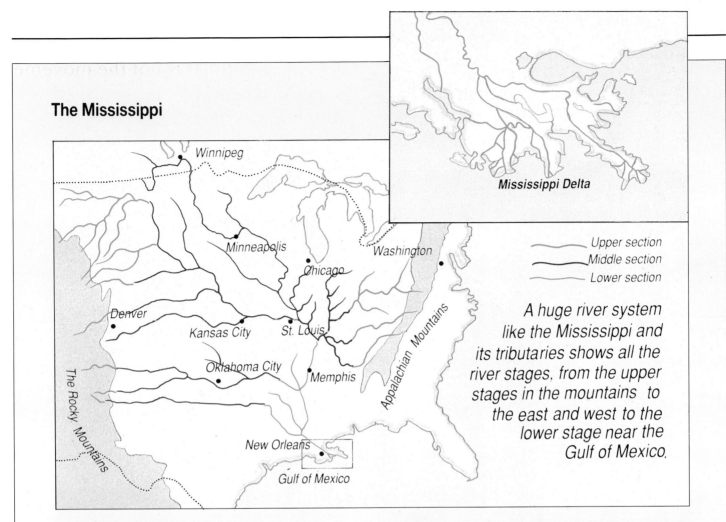

Mississippi Delta

Upper section
Middle section
Lower section

A huge river system like the Mississippi and its tributaries shows all the river stages, from the upper stages in the mountains to the east and west to the lower stage near the Gulf of Mexico.

The Mississippi and its tributaries form one of the greatest river systems of the world. The gorges of the Appalachians, and the eastern Rockies make up the upper river sections, while the great rivers of the Midwest – the Ohio, the Missouri and the Arkansas – make up the middle section of the system.

The rivers all flow into the Mississippi and near its mouth its lower section slows down and the material brought from all around is dropped. Here a great delta of islands and sandbanks is being built out into the Gulf of Mexico.

The meanders of the Missouri form as the river flows slowly over flat land.

Glaciers

Scoop up a handful of fluffy snow and squeeze it tightly into a snowball. It sticks together. This is because the pressure of your hands turns some of the snowflakes to ice, freezing the rest into a single mass.

The same thing happens in nature when snow falls upon snow and the lower layers have no chance to melt.

In mountain hollows, enormous weights of snow can pile up winter after winter, and eventually the bottom layers are squeezed into solid ice. Ice under pressure like this behaves quite differently from the brittle material that most of us know. When squeezed it becomes soft and flexible, like putty.

A mass of glacier ice creeps slowly downhill. The glacier scrapes debris from the valley walls and carries it along as dirty streaks of moraine.

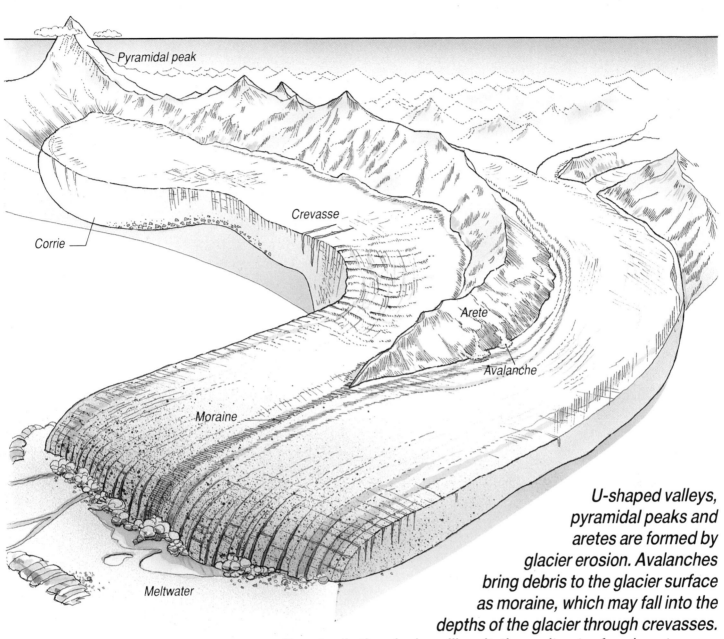

Pyramidal peak

Crevasse

Corrie

Arete

Avalanche

Moraine

Meltwater

U-shaped valleys, pyramidal peaks and aretes are formed by glacier erosion. Avalanches bring debris to the glacier surface as moraine, which may fall into the depths of the glacier through crevasses. Eventually the glacier will melt, the meltwater forming streams.

The solid river of ice moves slowly out of its hollow and creeps downhill. It becomes a **glacier**. More snow falls in the mountain hollow and the supply of ice is kept up. The weight of moving ice grinds out a U-shaped valley as it travels. Lumps of rock and gravel are torn up and carried along by it, sometimes forming dirty stripes along the glacier's surface. The surface of the ice is not under pressure and so it is brittle. The movement of the glacier cracks the ice and forms **crevasses** that may be very deep.

Eventually the glacier reaches lower ground and melts. The meltwater flows as rivers and streams from ice caves and crevasses. The meltwater is cloudy with ground-down rock fragments. Islands are formed from the rocky material brought down by the glacier. If the glacier is melting quickly it leaves behind mounds and ridges of rocks called **moraine**.

In places that are always cold, such as Greenland and Antarctica, glaciers can cover huge areas. Here the glaciers move outwards, rather than downhill, away from the build-up of snow and ice at the centre. Such glaciers are called ice-sheets.

An ice-sheet may cover a mountain range. Or it may squeeze between mountains and split up into a number of valley glaciers, possibly coming together again into another ice-sheet further on. The mountains sticking through the ice-sheet are called **nunataks**.

Meltwater streams are so full of debris that they produce great shingle-covered plains.

The Swiss Alps

Amongst the mountains of the Swiss Alps the Jungfrau is so high that the snow lasts all year. The build-up of snow produces a number of glaciers that flow down the valleys. One of the most famous is the Aletsch glacier, flowing southwards.

Smaller glaciers join the Aletsch from the east, bringing their own rocky moraine. The moraine lies as stripes along the glacier's length – stripes that are often the subject of tourists' photographs.

One valley to the east is dammed at its entrance by the glacier. It has filled with water to become a lake called the Marjelen See. The Aletsch eventually melts near the town of Belalp. The meltwater forms the River Massa that eventually flows down into the Rhône.

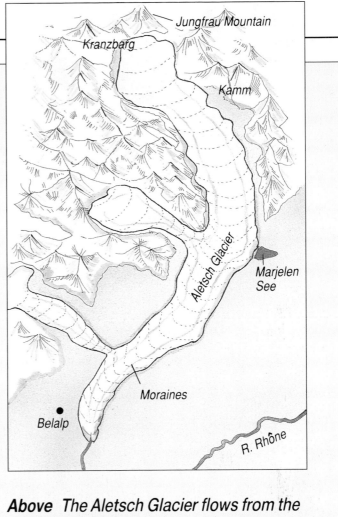

Above *The Aletsch Glacier flows from the Jungfrau. Its meltwater joins the Rhône.*

The northern oceans have icebergs, broken from the ends of glaciers.

When an ice-sheet reaches the sea, the movement of the waves and tides breaks it up. Huge chunks fall off and drift away as **icebergs** that are a danger to shipping. Over twenty thousand years ago, the last Ice Age ended. It spread ice-sheets over vast areas of North America, northern Europe and Asia. We can see this has happened because of the vast amounts of moraine that were left behind.

Weather

It is not the movement of continents, nor the pounding of the sea, nor the grinding of glaciers that produce the greatest changes to the landscape. It is the gentle rain and wind, the frost and ice, and all the other elements that we know of as weather.

Rain seeps into cracks and pores in the rocks and there it can freeze during cold spells. The ice expands and forces the rocks apart. Chunks of rock can shatter away from mountain faces in cold lands, and build up slopes of debris called **scree** at the foot of mountains.

Limestone caves are formed by underground streams. The minerals that are dissolved in the water build up as cone-shaped pinnacles on the floor and ceiling.

Frost-wedging process

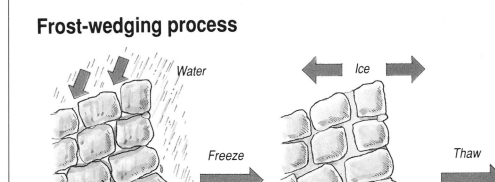

In cold, wet areas water seeps into the ground, finding its way into the cracks and joints in the rocks.

When it freezes the expansion of the freezing water in rock cracks is strong enough to wedge the cracks open. More water then seeps in, wedging the cracks open still further.

Eventually the cold rock splits into jagged lumps and falls away. The long slopes of broken rock called scree, found in cold areas, are a result of this frost-wedging process.

Rain can also dissolve the gas carbon dioxide from the atmosphere and become a weak acid. This can then break down the minerals in certain rocks like limestone and granite, and whole landscapes can crumble. Falling rain can move particles from rock and send them tumbling downhill, or wash them into rivers. All this action is called weathering.

Rocks are weakened by ground water passing through them and they are slowly broken up. Plants that grow on the surface of rocks send down their roots looking for water. When the roots meet the solid rock, they can force their way into cracks and break down the rock further. When the plants die, their remains are added to the debris. Burrowing animals like worms and moles tunnel through the mass, churning it up and mixing it thoroughly. The result is what we know as soil.

In dry areas wind-blown sand can blast rocks into all sorts of strange shapes.

Plant roots help to hold a soil together. In barren areas, where few plants grow, the soil tends to be very loose and dusty. Such areas are usually dry as well. With no moisture to stick the rock fragments together, the wind can blow them away. Sand is blown along close to the ground by the wind.

When a rock has wind-blown sand blasted against it, the result is like being worked on by sandpaper. The rock is polished smooth and worn away. Desert areas often have sandblasted features formed in this way. The eroding rock produces yet more sand. This adds to the mass of sand blown along in sandstorms.

The Dust Bowl

The Great Plains region of North America had such fertile grassland that the early settlers soon established farms there.

However, they grew the wrong sorts of crops, and the goodness of the soil became used up. When the plants died there were no roots to hold the soil together. It was now open to the effect of the weather.

Severe droughts in the 1930s turned the exposed soil to dust and blew it away. The rains that did come washed away the rest of the loose material. What was valuable farmland became a desert, called the Dust Bowl.

Many farmers abandoned their homes in the Dust Bowl in the 1930s as drought and overcultivation turned their fields to dust.

As we have seen, erosion is occurring in nature all the time. But people's activities can also cause erosion as well. In some parts of the world, such as Africa, overuse of the land by activities like farming can also affect the landscape.

Soil becomes eroded and is easily washed away when the rains come. This process of erosion eventually causes deserts to form.

When people's activities add to nature's influence in this way, the effects can be far-reaching.

Glossary

Beach   A gently-sloping shoreline made of sand or pebbles that are washed up by the sea.

Berm   A shelf of sand or shingle formed on a beach by the action of waves.

Block mountain   A mountain that is left standing as the landscape round about has fallen along faults.

Blowhole   A sea cave with a hole in the ceiling.

Continent   Any large mass of land and surrounding shallow sea.

Crevasse   A deep crack in the surface of a glacier.

Crust   The outermost layer of the Earth's structure.

Delta   This is formed when a river channel reaches the sea, and sandbanks and islands are built up by deposited material.

Dome mountains   Mountains made of granite, formed by hardened magma.

Estuary   A broad river mouth that can contain seawater at high tide and fresh water at low tide.

Fault   A crack in the rocks of the Earth's crust along which there has been some movement.

Fold mountains   Mountains formed when layers of rock at the Earth's crust were folded and compressed.

Glacier   A mass of ice in a valley, that moves slowly downhill.

Headland   A part of the coast that is higher than the land round about, and that sticks out into the sea.

Hot spot   An area beneath the Earth's crust that is hot, where volcanoes form.

Iceberg   A mass of ice broken from the end of a glacier that drifts in the sea.

Lagoon   An area of shallow water cut off from the sea by a sand bar or a coral reef.

Lava   Molten rocky material that erupts from a volcano.

Levée   A bank of sand and silt that is built up along the side of a river.

Magma   Underground molten rock.

Meander   A wide bend or a loop in a river.

Moraine   Rocky debris carried along by a glacier and deposited when the glacier melts.

Nunatak   A mountain that sticks up through an ice-sheet.

Scree   A slope made up of sharp fractured rocks at the base of a mountain.

Seastack   A column of rock sticking up out of the sea which was once part of a headland.

Spit   A bar of sand that extends part way across a river mouth.

Tectonic plates   Distinct sections of the Earth's surface that move in relation to one another.

Tributary   A small stream that flows into a larger stream or river.

Volcanic   To do with volcanoes. A volcano is a vent in the Earth's crust through which lava, steam and gases erupt and form a volcanic mountain.

Wave-cut platform   Gently sloping ledge at the edge of a cliff.

Books to read

Bender, L. *Glacier* (Franklin Watts, 1988)

Bender, L. *The Story of the Earth* (Franklin Watts, 1988)

Bramwell, M. *Mountains* (Franklin Watts, 1986)

Dineen, J. *Rivers and Lakes* (Macmillan, 1987)

Dixon, D. *The Planet Earth* (Franklin Watts, 1989)

Dudman, J. *Earthquake* (Wayland, 1992)

Dudman, J. *Volcano* (Wayland, 1992)

Jennings, T. *The Earth* (Oxford University Press, 1989)

Lye, K. *Coasts* (Wayland, 1988)

Lye, K. *The Earth* (Hamlyn, 1985)

Notes for activities

Copy an outline of the world from a map. Cut out paper shapes of the continents. Shuffle them together to see how South America and Africa once fitted together. Make Europe, Asia and India separate pieces and show how the Urals and Himalayas formed as they all joined together.

Scan books and newspaper articles for reports of volcanoes and earthquakes. Flag them on a map of the world. See if you can detect a pattern of where sections of the Earth's crust are slowly moving against each other.

Visit a beach if possible and identify any features caused by sea erosion and deposition. Take pictures of the beach in summer and winter. You may find that in winter high banks of shingle form, while in summer the beach is more sandy.

Find out how fast a river flows at different points along its bank. Record the time a floating object takes to flow along a selected distance. From this you can calculate the speed the river flows.

Fill a plastic bottle with water and allow it to freeze overnight. In the morning the top will have been forced off, or the bottle cracked because of the pressure of ice. This shows you how rocks in mountainous areas crack when the water in them freezes and expands.

Prepare a tray of soil in which grass is growing, and another tray where nothing is growing and stand them outside. Water the two trays and see how quickly the unplanted one washes away compared to the one with growing grass.

Index

DDs-93-325

Picture acknowledgements

The publishers would like to thank the following for allowing their photographs to be used in this book: Bruce Coleman Ltd 14 (Harold Lange), 21 (Nicholas de Vore), 24 (Charlie Ott); Geoscience Features 18, 26; Photri *title page*, 29; Frank Lane Picture Agency 10 (National Parks Service); Frank Spooner Pictures 13 (Aventurier/Loviny); Tony Stone Worldwide *cover* (Mike Surowiak), 4 (Tony Craddock), 5 (H. Richard Johnston), 6, 11 (Dennis Oda), 22, 25 (Geraldine Prentice), 28 (Robert Frerck); Wayland Picture Library *back cover*; Zefa Picture Library 8, 17 (Keith Klaus). Artwork is by Stephen Wheele.

Date Due